POISONOUS CREATURES

INVESTIGATE!

Beware the BOOMSLANG SNAKE!

Ursula Pang

Enslow PUBLISHING

Please visit our website, www.enslow.com. For a free color catalog of all our high-quality books, call toll free 1-800-398-2504 or fax 1-877-980-4454.

Library of Congress Cataloging-in-Publication Data

Names: Pang, Ursula, author.
Title: Beware the boomslang snake! / Ursula Pang.
Other titles: Poisonous creatures.
Description: New York : Enslow Publishing, 2023. | Series: Poisonous creatures | Includes index.
Identifiers: LCCN 2021048987 (print) | LCCN 2021048988 (ebook) | ISBN 9781978527362 (library binding) | ISBN 9781978527348 (paperback) | ISBN 9781978527355 (6 pack) | ISBN 9781978527379 (ebook)
Subjects: LCSH: Boomslang–Juvenile literature. | Boomslang–Venom–Juvenile literature. | Poisonous snakes–Africa–Juvenile literature.
Classification: LCC QL666.O636 P36 2023 (print) | LCC QL666.O636 (ebook) | DDC 597.96/165–dc23/eng/20211008
LC record available at https://lccn.loc.gov/2021048987
LC ebook record available at https://lccn.loc.gov/2021048988

Portions of this work were originally authored by Alicia Z. Klepeis and published as *The Boomslang Snake.* All new material in this edition was authored by Ursula Pang.

Published in 2023 by
Enslow Publishing
29 E. 21st Street
New York, NY 10010

Designer: Katelyn E. Reynolds
Interior Layout: Tanya Dellaccio
Editor: Greg Roza

Photo credits: Series background Vector Tradition/Shutterstock.com; cover (snake) CHAKRart/Shutterstock.com; p. 5 (top) Terrence L'Estrange/Shutterstock.com; pp. 5 (bottom), 7, 9, 12, 17 (bottom), 23 (top) Willem Van Zyl/Shutterstock.com; p. 6 Milan Zygmunt/Shutterstock.com; p. 8 reptiles4all/Shutterstock.com; pp. 10, 19 Stu Porter/Shutterstock.com; p. 11 NickEvansKZN/Shutterstock.com; p. 13 Sprocky/Shutterstock.com; p. 17 (top) mbrand85/Shutterstock.com; p. 18 Stephanie Periquet/Shutterstock.com; p. 21 (top) Evelyn D. Harrison/Shutterstock.com; p. 21 (middle) Ken Griffiths/Shutterstock.com; p. 21 (bottom-right) jaroslava V/Shutterstock.com; p. 23 (bottom) Afropreneur/Shutterstock.com; pp. 24, 28 Eugene Troskie/Shutterstock.com; p. 25 valon.red/Alamy Stock Photo; p. 27 Ton Bangkeaw/Shutterstock.com; p. 29 Martin Knoedel/Shutterstock.com.

Printed in the United States of America

CPSIA compliance information: Batch #CSENS23: For further information contact Enslow Publishing, New York, New York, at 1-800-398-2504.

CONTENTS

WORDS IN THE GLOSSARY APPEAR IN **BOLD** TYPE THE FIRST TIME THEY ARE USED IN THE TEXT.

A DEADLY DISCOVERY

In 1957, snake expert Karl Schmidt was working at the Field Museum in Chicago. While examining a young boomslang snake, he was bitten on his thumb. At the time, rear-fanged snakes were not considered dangerous. Schmidt didn't seek treatment, but he wrote down his **symptoms**. The next day, he died from the snake's venom.

From its appearance to its venom, the boomslang snake is one amazing animal. These snakes are arboreal, which means they make their homes in trees and shrubs. They live in Africa, south of the Sahara.

Boomslangs can thrive in different **habitats**. They live in savannas and lowland forests. They also live in grasslands and semidesert areas known as karoos.

TREE-CLIMBING HUNTER

Boomslang snakes are active during the day, but they are excellent at hiding. They spend most of their time in the trees. They slither along branches until they find a good hiding place. They will slither to the ground for food or to **bask** in the sun.

Large meat-eating birds eat boomslangs. Ospreys and secretary birds hunt them. So do falcons, kestrels, and other raptors.

Boomslangs are carnivores, or meat eaters. Their diet is made up mainly of small tree-dwelling lizards and frogs. They often eat chameleons. Sometimes they eat small mammals, birds, and the eggs of birds and reptiles. Boomslangs have also been known to eat other boomslangs.

SIZE AND COLOR

Boomslangs are small snakes. Adults are usually between 3 and 5 feet (0.9 to 1.5 m) long. In some parts of their range, they can be more than 6 feet (1.8 m) long. These snakes have thin bodies. At birth, boomslang **hatchlings** are about 8 inches (20 cm) long. Boomslang snakes have small, egg-shaped heads.

BOOMSLANG HATCHLINGS ARE SKINNY WITH BRIGHT GREEN EYES THAT TAKE UP MUCH OF THEIR HEAD.

Get the Facts!

Young boomslangs are more colorful than adults. They often have a stripe down their back. This stripe usually fades as they become adults.

Not all boomslangs are the same color. Females are usually greenish-brown with light brown bellies. Males, however, can be more colorful. They can be dark green, bright green, dark brown, or black. They may have bright yellow bellies.

BOOMSLANG EGGS

The breeding season for boomslang snakes lasts from July to early October. During this time, male boomslangs may fight each other. They fight to gain the right to **mate** with a female. Boomslang snakes are oviparous. That means they lay eggs.

A female will usually lay between 8 and 27 eggs. The eggs have soft shells. They are about the size of ping-pong balls. In about 65 to 100 days, the eggs will hatch.

BOOMSLANGS LAY THEIR EGGS IN DAMP PLACES, LIKE A TREE HOLLOW, SO THEY DON'T DRY OUT. THE MOTHER DOES NOT STAY AROUND TO CARE FOR THE EGGS.

Get the Facts!

Boomslangs have large eyes for snakes, and they have excellent eyesight.

GROWING UP

Boomslang snakes are small. However, they are deadly from the time they are born. Even hatchlings can deliver **lethal** doses of venom. This keeps them safe from most predators, unless another boomslang or a big meat-eating bird attacks.

YOUNG BOOMSLANGS HIDE IN TREES AND TALL GRASS IN ORDER TO SURPRISE THEIR **PREY.**

Get the Facts!

Baby boomslangs need to eat once every two to three days. Their diet is mostly smaller reptiles.

Boomslang hatchlings are on their own. They are shy and try to stay away from danger. They will shed their skin for the first time within 10 days of hatching. On average, boomslangs live about eight years in the wild.

GETTING TO KNOW THE BOOMSLANG SNAKE

Scientific Name

Dispholidus typus

Common Name

"Boomslang" in the Afrikaans language of Southern Africa means "tree snake."

Size

Usually 3 to 5 feet (0.9 to 1.5 m) as adults but can be more than 6 feet (1.8 m) long.

Coloration

The boomslang is one of the few snakes in which color is different between males and females. Females usually have dark and dull colors. Males often have brighter colors.

Deadly Venom

The boomslang carries up to 8 milligrams (0.0016 teaspoons) of venom.

Big Bite

The boomslang delivers venom with fangs (3 to 5 millimeters, or less than 0.2 inches long) located in the back of its mouth. It can open its mouth 170 degrees so it can bite arms and legs.

AFRICA

THIS MAP SHOWS THE RANGE OF THE BOOMSLANG IN GREEN.

BORN TO HUNT

Boomslangs are born hunters. They have large eyes and excellent eyesight. Like many snakes, they have a forked tongue that can sense matter in the air.

Boomslang coloring allows them to blend in with trees, shrubs, and long grasses. They sit very quietly and remain very still until prey wanders by. They begin to move their head from side to side, **mimicking** the movement of leaves or branches swaying in the wind. The prey thinks the snake is just a plant! The boomslang strikes quickly, and the prey rarely sees it coming.

ONCE A BOOMSLANG ATTACKS, ITS SHARP FANGS **INJECT** DEADLY VENOM INTO THEIR VICTIMS.

Get the Facts!

When a boomslang senses danger, it fills its upper neck with air. This makes the snake look bigger and shows the bright skin between its scales.

BOOMSLANG VENOM

Boomslangs are rear-fanged snakes. The boomslang's fangs are longer than most rear-fanged snakes. The venom runs down grooves in the fangs. A boomslang may strike a few times. This makes sure enough venom is injected into its prey.

Venomous animals inject venom into their prey. They use fangs, stingers, or spines. Venom is different from poison. Poisonous animals make toxins that are harmful when eaten or touched.

BOOMSLANG SNAKES HIDE UNTIL PREY SHOWS UP AND THEN STRIKE SUDDENLY.

Boomslang venom doesn't act fast. It can take several hours—even up to a day—for the effects of the venom to show up in a person. It can cause bleeding, headaches, and sleepiness, among other issues.

THE WORLD'S DEADLIEST SNAKE VENOMS

Scientists use a test called LD50 to rank the deadliest snake venoms in the world. "LD" stands for "lethal dose," and "50" stands for "50 percent" of the test subjects. Animals, usually mice, are injected with venom. The amount of venom that kills half the mice results in the LD50 number. The lower the LD50 number, the more deadly the venom.

How much boomslang venom is needed to kill a person? Five milligrams of venom could kill a human. That's 0.001 of a teaspoon!

TIGER RATTLESNAKE

SCIENTISTS ARE FINDING NEW WAYS TO TEST SNAKE VENOM TO STOP HARMING MICE AND OTHER ANIMALS.

SNAKE	LD50 (MG/KG)	VENOM IN A BITE (MG)
HOOK-NOSED SEA SNAKE	0.02	7 TO 79
RUSSEL'S VIPER	0.03	130 TO 250
INLAND TAIPAN	0.03	44 TO 110
DUBOIS'S SEA SNAKE	0.04	0.7
EASTERN BROWN SNAKE	0.05	2 TO 67
BLACK MAMBA	0.05	50 TO 100
TIGER RATTLESNAKE	0.06	6 TO 11
BOOMSLANG	0.07	1.6 TO 8
YELLOW-BELLIED SEA SNAKE	0.07	1 TO 4
COMMON INDIAN KRAIT	0.09	8 TO 20

INLAND TAIPAN

RUSSEL'S VIPER

BLOOD TROUBLE

Boomslangs make venom in glands that **evolved** from salivary glands. These are the glands humans use to make spit. The venom is stored in a sac inside the snake's head. It stays there until the snake bites its prey.

Boomslang venom, a hemotoxin, affects the blood of its victims. It causes red blood cells to break open, resulting in bleeding inside the body. Hemotoxins can also stop the blood from clotting. When blood clots, it gets thick and bleeding stops. Furthermore, hemotoxins can also harm an animal's tissues and organs.

Get the Facts!

Boomslangs fear people and keep away from them. They can sense a person long before a person can see them. When you approach a boomslang, it will get away before you even know it's there.

MOST ATTACKS BY BOOMSLANGS ONLY HAPPEN WHEN PEOPLE TRY TO CAPTURE OR KILL THEM.

SLOW DEATH

Boomslang venom acts slowly. Because of this, victims may not think they need to rush for medical help. Symptoms may start with a headache and nausea, followed by sleepiness. However, the hemotoxin spreads throughout the body and causes the victim to bleed to death.

People who were bitten can bleed from their nostrils and gums. Their will be blood in their pee and poop. Some peoples' bodies look bluish because of all the bleeding inside their body. Many also say they see yellow. This could be due to bleeding inside their eyes.

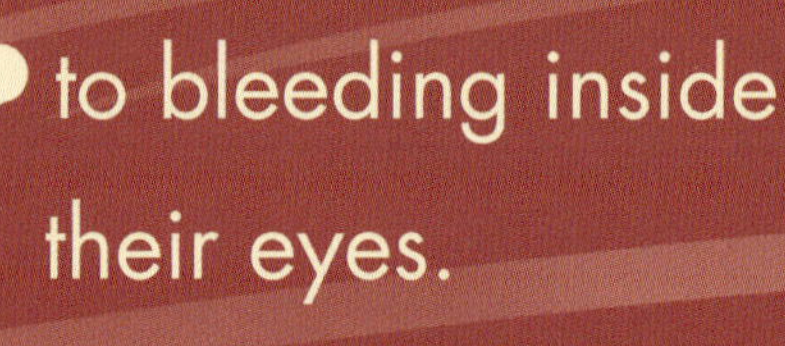

IT'S A GOOD THING THAT BOOMSLANGS ARE SHY. FEWER THAN 10 REPORTED DEATHS FROM THEIR BITES HAVE BEEN RECORDED WORLDWIDE.

Get the Facts!

Scientists created boomslang antivenom (also spelled antivenin) in the 1940s. An antivenom is a drug made from the snake's venom itself. If given on time, it can counteract the effects of the venom.

MAKING ANTIVENOM

Boomslang venom can be useful to scientists. The collected venom is used to make an antivenom that saves lives every year. This antivenom is the only known way to cure a bite victim. Giving the patient fresh blood and **plasma** is also helpful to recovery.

South African scientists have been important in the study of boomslangs and their venom. Future research may make it cost less to produce boomslang antivenom. This could help make it more available to Africans who live in rural areas where the snakes live.

SNAKE HANDLERS HAVE A VERY DANGEROUS JOB. WHEN GATHERING SNAKE VENOM, THEY MUST KEEP THE SNAKE FROM TURNING ITS HEAD AND BITING.

Get the Facts!

Specially trained snake handlers carefully grab the back of the boomslang's head. While in this position, they can press on the snake's venom glands. The venom gets squeezed into a container. It takes the venom of many snakes to create antivenom.

ESSENTIAL TO THE ECOSYSTEM

Boomslangs frighten people all over Africa. They're known as one of the most venomous snakes in the world. But do they have any benefits for people? Most scientists would say they do. All animals are important to the habitats where they live. This is even true for venomous ones.

Boomslang populations, or numbers, seem to be healthy throughout their range.

BOOMSLANGS MIGHT BE DEADLY, BUT THEY ARE VERY IMPORTANT TO THE HEALTH OF THE PLACES IN WHICH THEY LIVE.

Boomslangs are part of the food web. They eat lizards, birds, and other small animals. Some of the rodents they eat can harm crops. These rodents also carry illnesses, so reducing their numbers is good for farmers and others who live nearby.

GLOSSARY

bask To lie exposed to light and warmth, usually from the sun.

evolve To change slowly into a state that is more advanced.

habitat The natural place where an animal lives.

hatchling A recently hatched animal, such as a snake.

inject To force a liquid into something using a sharp, hollow tool, such as needle, fang, or stinger.

lethal Deadly.

mate To come together to produce babies

mimic To copy the look or behavior of something else.

plasma The colorless fluid in blood that carries blood cells.

prey An animal hunted or killed by another animal for food. Also, to hunt other animals for food.

symptom A noticeable change in the body or its functions that indicates the presence of an illness.

FOR MORE INFORMATION

Books

Murray, Julie. *Boomslangs*. North Mankato, MN: Abdo Publishing, 2017.

Taylor, Barbara. *100 Facts—Snakes*. Thaxted, UK: Miles Kelly, 2019.

Websites

Boomslang

www.africansnakebiteinstitute.com/snake/boomslang/

The African Snakebite Institute offers more information about the boomslang snake, including colorful photographs, a video, and links to other African snakes.

Wildlife Guide to the Boomslang

safarisafricana.com/animals/boomslang/

Read more about the boomslang and see more colorful photographs of them. Also includes a video of a boomslang hunting a chameleon.

Publisher's note to educators and parents: Our editors have carefully reviewed these websites to ensure that they are suitable for students. Many websites change frequently, however, and we cannot guarantee that a site's future contents will continue to meet our high standards of quality and educational value. Be advised that students should be closely supervised whenever they access the internet.

INDEX